FLOWERS AND THEIR VISITORS

BLACK'S PICTURE INFORMATION BOOKS

Scientific Adviser Jean Imrie M Sc, Area Adviser, West Riding CC

flowers and their visitors

JANET DAVIDSON

Adam and Charles Black · London

Published by A & C Black Ltd
4, 5 & 6 Soho Square, London W1V 6AD

©1970 Moussault's Uitgeverij NV, Amsterdam
©1973 (English edition) A & C Black Ltd

ISBN 0 7136 1316 5

First published in the Netherlands by Moussault's
Uitgeverij NV, under the title *Bloemen en hun
bezoekers* with a text by G den Hoed; this text
translated by Adrienne Dixon and adapted by Janet
Davidson with the assistance of Jean Imrie.

Filmset by Photoprint Plates Ltd, Rayleigh, Essex
Printed in the Netherlands by Ysel Press, Deventer

Acknowledgements

The colour illustrations in this book are reproduced
by courtesy of C and H Rol and J Voerman, Jr.
They have been taken from the book *De bloemen en
haar vrienden*.

The black and white illustration on page 8 is by C
Rol; the other line drawings are by G den Hoed.

Contents

Foreword	**7**
Reproduction by means of flowers	**9**
Fertilisation	9
Stamens and pistils	9
Pollination	9
Transport of pollen	11
The animals which visit plants	11
Insects which visit plants	11
Supply and demand	12
'Advertising'	12
The visitors who profit	**13**
How long is their tongue?	13
Regular visitors	13
Regular accidental visitors	14
Occasional visitors	14
Harmful visitors	15
Bees, nectar and honey	15
Pollen collectors	16
Pictures and descriptions	**17**
Survey of insect-pollinated flowers	**41**
Classification	41
Tongue length of visitors	41
Roses	42
Poppies	42
Umbellifers	43
Lime	43
Willow	44
Swede	44
Coltsfoot	44
Ling	45
Forget-me-not	45
Birthwort and lords-and-ladies	46
Grass of Parnassus	46
Cowslip	47
Purple loosestrife	47
Willow-herb	48
Common figwort	48
Monkshood	48
Papilionaceous flowers	49
Labiate flowers	50
Wild honeysuckle	50
Yellow flag	51
Orchids	51
Glossary	**52**
Other books about flowers and insects	**53**
Index of Latin names	**54**
Index	**55**

Many flowers depend on insects if they are to reproduce themselves. The insect which visits a flower may carry pollen from it to the flower of another plant of the same species. This results in pollination, leading to fertilisation and to new seed.

Insects only visit flowers out of self-interest. The pollen of some flowers is edible, and insects come to collect it. In many cases flowers produce nectar, a sweet liquid which insects enjoy. Unlike pollen, nectar has nothing directly to do with the reproduction of the plant, but is produced in order to attract insect visitors. The same is true of the scent of some flowers, and even their shape and colour may help insects to identify sources of food.

The shape of flowers, and the shape of insects, has developed over many millions of years by the process of evolution. This has been expressed in very simple terms as 'the survival of the fittest'. Occasionally a plant grows which is slightly different from the others of the same species—perhaps by having one slightly brighter petal. If insects can spot that flower more easily than others, they are more likely to visit it, so it has a better than average chance of reproducing itself. In this way an 'improvement' may gradually spread itself throughout the species.

Some of the plants you will find in this book have evolved a very special shape. Some only attract or can accommodate one type of visitor, for example the monkshood which would probably die out if bumblebees were to die out.

Some have never given up other methods of reproduction, for example, the lesser celandine which never produces seed at all (it reproduces by underground tubers) but still produces a great deal of pollen.

At the same time you will find many examples here of flowers with 'mechanisms' for placing pollen on the body of an insect and then on to the stigmas of flowers, which we would say were examples of brilliant engineering if man had invented them.

White deadnettle

We do not know what the first flowers looked like, millions of years ago, but we think that they were probably bud-like clusters of ordinary green leaves grouped together at the top of the stalk.

After an incredibly long evolution, flowers have developed in many different ways, each plant making sure that its species can survive. This survival is only possible if enough seeds germinate, and grow into new plants.

Fertilisation

Flowering plants can be male, female or bisexual (that is, male and female at the same time). Usually they are bisexual. The male sexual organs are the *stamens* which produce male cells (*pollen*). The female sexual organs are the *pistils* which produce female cells (*ovules*). Each ovule can only be fertilised by one grain of pollen.

All plants must find a solution to the same problem—how to get the pollen to the ovule.

carpels
stamens
corolla
sepals
leaves

A very primitive flower

Stamens and pistils

The grains of pollen develop inside the part of the stamen called the *anther*. Usually two *pollen sacs* make up each anther, and they take many different forms.

The ovules usually develop inside *ovaries*, which are part of the pistil. Within the ovary are one or more chambers which hold the ovules. After fertilisation, the ovary grows into a *fruit*, and the ovule grows into a *seed*.

The part of the pistil which is to receive the pollen grains is called the *stigma*. When pollen grains reach a mature stigma, they put out hair-like tubes, which as they grow, make their way towards the ovules. Sometimes the stigma is separated from the ovary by a *style*.

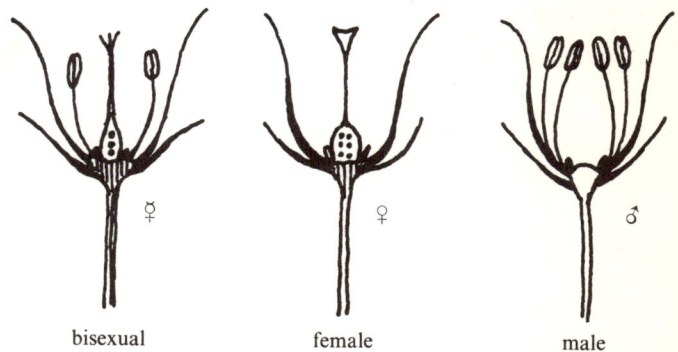

bisexual   female   male

Pollination

When grains of pollen reach the stigma, they remain there because it is sticky. But what kind of pollen is it? If it is pollen from a plant of a different species, nothing may happen.

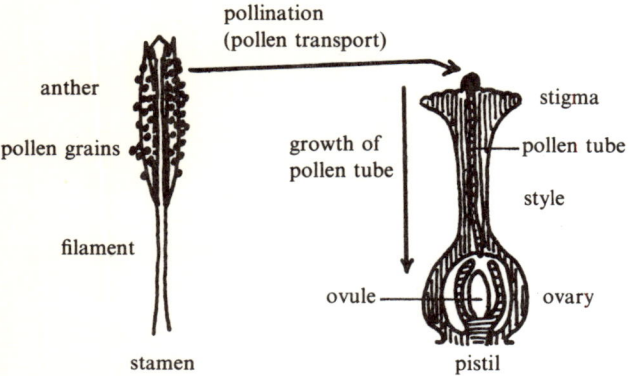

Pollination and fertilisation

Before the pollen tubes can grow towards the ovules, they need certain substances produced from the stigma. Only pollen from the same species, or from closely related species, usually react to these substances.

If the pollen comes not from the same but from a related species, the new plant produced will be a *hybrid*; gardeners would call it a 'cross'.

Where the flower is bisexual, the pollen may have come from the stamens of the same flower or from another flower on the same plant and this is called *self-pollination*. If the pollen comes from the flower of another plant, this is called *cross-pollination*.

In general we can say that cross-pollination is best, because it allows for the possibility of slight changes in the species. As a result, a plant may develop some minute oddity. This oddity may be inherited or even exaggerated in the next generation. If the oddity is of benefit to the plant, it will persist, but if it is not, the plants may not survive.

Most plants make provision for cross-pollination. A few are always self-pollinated. Some flowers are as sterile to their own pollen as they are to pollen from plants of another species. Others provide for self-pollination if cross-pollination fails.

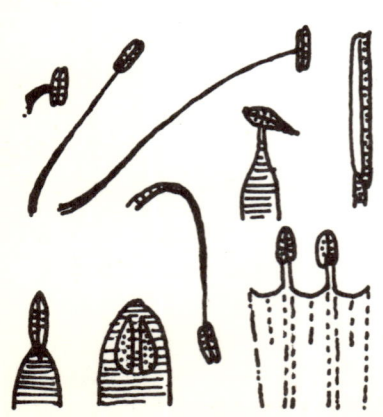

Various forms of stamen

Various forms of pistil

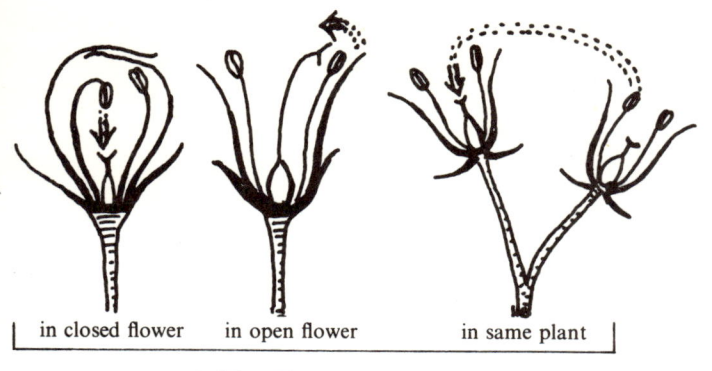

in closed flower   in open flower   in same plant

Self pollination

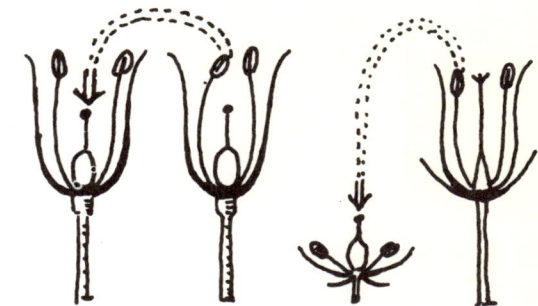

Cross pollination between plants of same species

Pollination between plants of different species

Transport of pollen

The simplest way to achieve pollination is for the pollen to fall on to the stigma in the same flower, or one near by. To give a slightly better chance of cross-pollination, some plants give the pollen a 'push'. This happens, for example, when the flowers of the birds'-foot trefoil are visited.

A very few water plants depend upon water for pollination, but pollination by wind is very common. We can assume that the most primitive flowers were wind pollinated.

Many flowering plants rely upon animals to spread their pollen. Animal pollination is only possible where animals visit the plant and of course they only do this out of self-interest. The plants which can attract animals by offering them something to eat are at an advantage, and some species may have died out because they were unable to compete.

The animals which visit plants

The animals which visit plants can be divided as follows:

a. A few birds (especially humming birds), mammals (such as small rodents), and some species of bat visit plants to eat the flowers, nectar or pollen. No examples of pollination by this method are found in Britain.

b. Slugs and snails, which crawl slowly over the flowers. This may occur in the case of some marsh plants, but it is usually accidental and no species relies on this method.

c. Many kinds of insect. Without these insect visitors, many plants could not survive.

Insects which visit flowers

Some insects visit flowers 'on foot', for example, worker ants, but most insects fly from flower to flower. This helps the plant by making cross-fertilisation more likely. The best carriers of pollen are flying insects such as gnats, flies, wasps, bees and butterflies.

We can immediately divide these insects into two large groups:

a. Those whose larvae depend on nectar and pollen collected for them by the adults.

b. Those whose larvae do not depend on this.

In the second group we find some insects which are regular visitors, and some which visit only occasionally. Group (a) contains many insects which could not survive without flowers.

During the millions of years of evolution, both the flowers and the insects have adapted themselves to each other. Certain flowers suit certain insects and vice versa.

Supply and demand

Plants can build up food from carbon dioxide and water, using the sun's energy. Animals cannot do this, and they depend on the plants for food. These plant foods include nectar and pollen as well as leaves, fruits, seeds, etc.

Nectar is a watery solution of sugars, which is only produced by the plant to attract animals (mainly insects).

If the animals eat the pollen, the chance of pollination is less, so many flowers with edible pollen produce it in large quantities to allow for some to be eaten. A few flowers produce no nectar at all, and rely on greater quantities of edible pollen.

'Advertising'

Flying insects have large compound eyes made up of many smaller facets by means of which they can distinguish different shapes and, often, different colours. Many insects also respond to different scents, and they show preferences for the taste of certain kinds of nectar.

Animals can learn from experience, and insects learn that a particular shape, colour or smell means a particular kind of nectar or pollen.

The visitors who profit

How long is their tongue?

An insect can only reach the nectar a flower produces if its *proboscis* or 'tongue' is long enough. Most beetles, many flies and a few bees have short probosces, so they only visit flowers where the nectar is near the surface.

Some flies, many bees and almost all butterflies have long probosces, so they can reach nectar which is deeper, hidden within the flower. A few bees and various butterflies have very long probosces and can extract the most deeply seated nectar.

Insects with very long probosces can visit all flowers, even those where the nectar is very near the surface. For example, the nectar in the flowers of the buddleia (see page 25), is not deep down, but butterflies are attracted to it as well as insects with short probosces.

Insects with short probosces can obtain nectar from a few flowers only and usually they do not rely on this visiting for the whole of their food. Sometimes a flower drips nectar, for

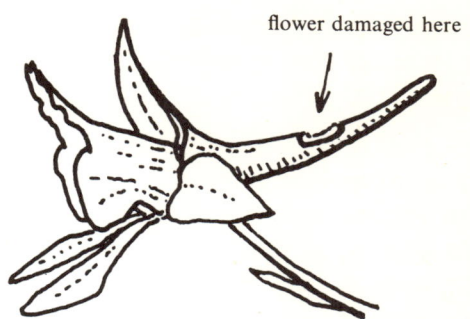

Some bumblebees bite into the base of the flower —in this case a larkspur

example the lime, and then these insects can readily feed upon it.

Some bumblebees bite through the base of a flower, for example the flowers of white dead nettle and larkspur, to get at the nectar. This forced entry does not benefit the plant as the insect avoids the stamens and the pistils.

Regular visitors

These include the butterflies, bees and bumblebees. Butterflies only visit the flowers as adult insects. (Their caterpillars are leaf-eaters and damage other parts of plants). Since the adult insect does not grow in size, it needs only a small amount of liquid and some food which it can quickly digest. The butterfly dips its long, tube-like proboscis into the flower, and there is little chance of it picking up any pollen at the same time.

Adult honeybees also visit flowers but they need a great deal of nectar and pollen for their larvae and the non-workers in the hive. They must also lay in stores for the winter, so the

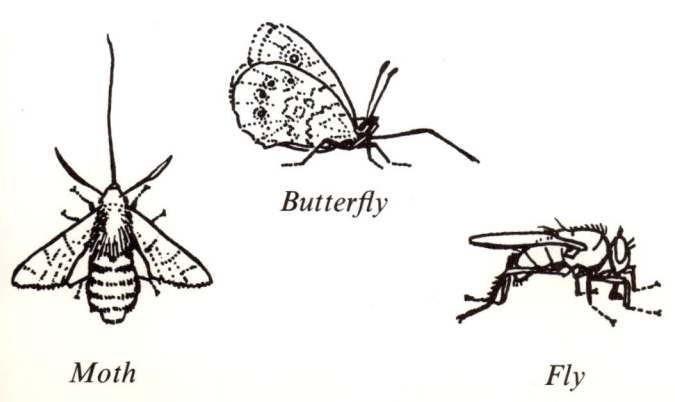

Butterfly

Moth

Fly

workers have a great deal to do, and need large amounts of food to supply their own energy.

Bumblebees also live in communities, and build up stores of honey, but the whole hive, except the queens, dies before winter, so there is not the same feverish activity as with the honeybees.

Regular accidental visitors

These include many flies and gnats, and many species of wasp including the common wasp. They all have short or medium-length tongues, and visit flowers where the nectar is near the surface.

Yet there are flowers which specialise in this kind of visitor, for example *Parnassia palustris* (page 30), visited by hoverflies; lords-and-ladies (page 39) and birthwort (page 43), visited by much smaller flies; marsh helleborine (page 37) and flowering rush (page 24) which are pollinated by wasps.

Adult wasps need little food and most of them do not live in communities. One notable exception is the common wasp. Of those species which do live in communities, most will eat anything; they may visit the flower to suck nectar, but they may well visit only to attack and kill another visitor.

Occasional visitors

These are the beetles and bugs. They have no community to feed, and they do not look after their larvae, but some species are always

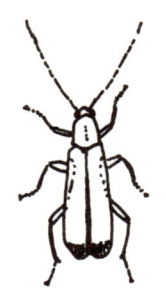

Soldier beetle

Ladybird

Long-horn beetle

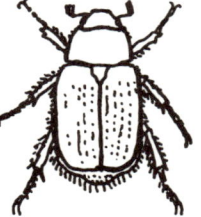

Leaf beetle

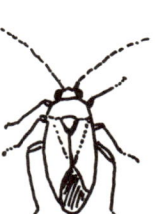

Flower bug

Bumblebee

Hoverfly

Common wasp

Owl midge

Harlequin fly

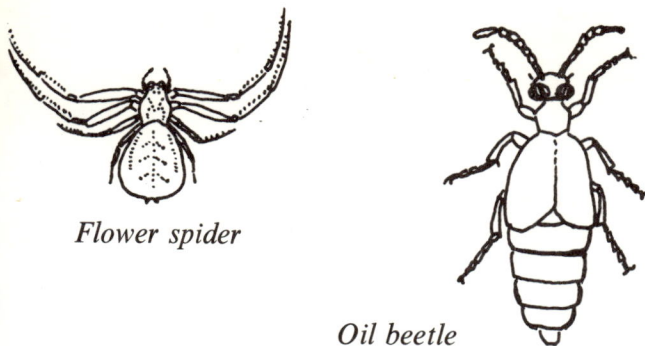

Flower spider

Oil beetle

found on the flowers of the *Umbelliferae* family.

The garden chafer (page 22) visits roses to eat the petals. It may occasionally cause pollination but probably does more harm than good.

Harmful visitors

We have already mentioned the garden chafer and the bumblebee as visitors which are sometimes harmful. Wasps can make pollination less likely by killing other visitors. So do some spiders; they may spin a web round or under the flowers, and spring forward the moment an insect touches the web. Other kinds of spider actually live inside the flower, using it as a trap. Some of them are the same colour as the flower, for camouflage.

Bees, nectar and honey

Bees and bumblebees suck up the sugar solution (nectar) from flowers and collect it in their 'honey stomach', which is a pouch, separate from the digestive system. There is a valve between the two, and the bee is able to

transfer the nectar into its digestive system when it needs it for its own food.

In the honey stomach, the nectar is changed into honey by special juices. What the worker bee does not itself need, it disgorges when it returns to the hive. Part of the honey is used to feed the non-workers, part of it is stored for the winter, and part of it is put in the breeding cells or 'comb' for the larvae to eat. The cell walls are made of beeswax which is secreted by the bees. Beeswax is not made from nectar or pollen.

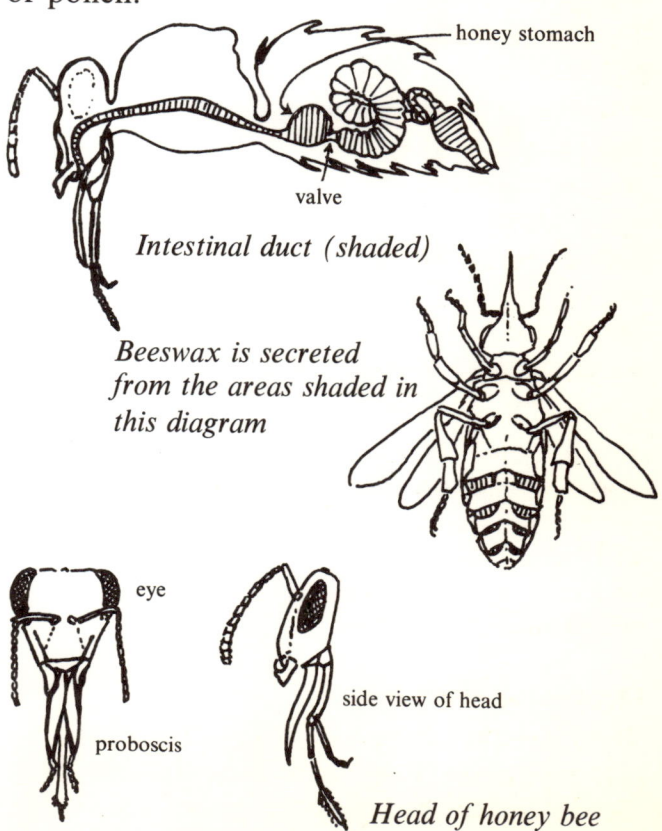

Intestinal duct (shaded)

Beeswax is secreted from the areas shaded in this diagram

Head of honey bee

The honey keeps the flavour of the flowers from which it was made, so when we take the honey we can label it heather honey, lime honey, clover honey etc. The beekeeper takes honey out of the hive and replaces it with sugar, on which the bees feed during the winter.

Pollen collectors

When bees and bumblebees collect pollen, it sticks on to their bodies. Then they carry it to their hives or nests. Some small wild bees can carry a small amount of pollen in their mouths to their nest, where, at each visit, they fill a cell with food and deposit an egg there.

Other wild bees have thick hairs on the under-side of their bodies and on their legs. If they move about over the anthers for some time, a lot of pollen sticks in the hair. Once in the nest, this is 'combed out' by means of stiff bristle-like hairs on the legs, and left as food for the larvae.

Some wild bees have brush-like hairs on their abdomen, in which the pollen accumulates.

Honey-bees and bumblebees (at least, the females and workers) have special hollows on

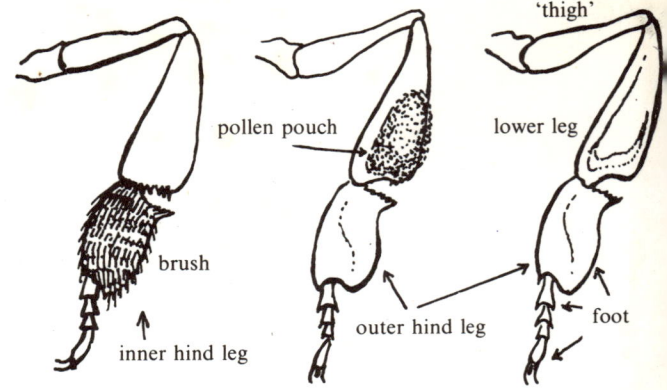

The hind legs of honey-bees and Bumblebees

the outside of their hind legs in which they can carry pollen. As each worker flies out of the nest, he takes along some honey as a 'glue'. Inside the flower the bee rubs its head and fore-legs on the pollen, rolling it into balls stuck together with honey. During flight, these balls are passed to the rear legs, using the many stiff bristles on the legs and feet.

Back in the nest, the bee pushes the pollen out of the pouches on its legs, and passes it to other workers from the same colony, which put the pollen into cells within the hive. The store of pollen, like the store of honey, is used as needed by both larvae and adult bees.

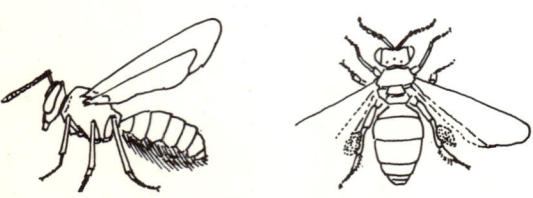

Left: abdominal hairs; right: hind legs with pollen 'baskets'

Pictures and descriptions

1. Ribwort plantain *Plantago lanceolata* **with sun fly**

This plant relies on wind pollination, having fairly dry pollen and no nectaries, but it is nevertheless visited and pollinated by insects too.

The inflorescence has inconspicuous flowers; each little flower has four stamens and a protruding stigma. The pistils ripen before the stamens, which is one way of avoiding self-pollination.

2. Coltsfoot *Tussilago farfara* **with small tortoiseshell**

What looks like a single flower is in fact an inflorescence, with many tiny flowers or 'florets' at the top of one stalk. It is a typical insect-pollinated plant, offering nectar and having a distinctive shape. It flowers early, before the leaves appear, which makes its flowers easier to see. The small tortoiseshell is a frequent visitor, because it survives the winter. See also page 44.

3. Cornflower *Centaurea cyanus* **with red-tailed bumblebee**

A cornflower inflorescence is a head of tiny florets surrounded by scale leaves. In the centre there are tube-shaped florets, and on the outside funnel-shaped florets. These outer florets are conspicuous and attract insects, but have no stamens or pistils. In the inner florets, five stamens form a cylinder, and the unripe pistil grows up through the cylinder, pushing the pollen upwards.

4

4. Goat willow *Salix caprea* with mining bee

Willow catkins are inflorescences which contain either all male or all female flowers. Each tree is either male or female.

Most trees with catkins depend upon wind for pollination, but willows rely on insect visitors. In sunny spring weather the flowers (both male and female) produce nectar and they are easily seen because the leaves have not yet unfurled. See also page 44.

THE INSECTS IN PICTURES 1–4

1. Sun fly *Helophilus pendulus*

Among the many species of hover-flies, quite a number have yellow and black markings, so that they are often mistaken for wasps. (Wasps have four wings, hoverflies only two.) Hoverflies can change direction rapidly between forward flight and hovering over one spot.

2. Small tortoiseshell butterfly *Vanessa urticae*

The final generation of each year survives the winter in large numbers. These butterflies can be found from early spring until late summer.

3. Red-tailed bumblebee *Bombus lapidarius*

Bumblebees are recognisable by the red colour on their abdomen. Any bumblebee found in early spring will be a queen which has hibernated after fertilisation the previous September. They start making a nest about May, and the community will include workers, males and females.

4. Mining bee *Andrena haemorrhoa*

The female is covered with dense reddish hair, and the male with white hair. The nest is made in sandy soil.

5. White clover *Trifolium repens* **with early bumblebee** *Bombus pratorum*

This plant is common in fields and gardens. The flowers are at first white or pink, and attract insects, but later they turn brown and are no longer visited. The nectaries are at the base of the flower, so only insects with long probosces can reach them. They are often visited by the early bumblebees. See also page 49.

6. Red clover *Trifolium pratense* **with large copper butterfly** *Lycaena dispar*

This kind of clover is found by the roadside, and is also specially sown in fields as food for animals. The flowers are very long and narrow, so only insects with very long tongues can reach the nectar — butterflies and some long-tongued bumblebees. The butterfly illustrated, the large copper butterfly, now extinct in Great Britain, is being reintroduced.

HOW THE MECHANISM WORKS

Various flowers, including the clovers, gorse, broom, peas, beans, sweet peas and laburnum have a special mechanism which makes cross-pollination more likely.

A clover flower has five petals. The upper petal which is broad and conspicuous is called the 'standard'. There are two shorter 'wing' petals, and a 'keel', which consists of the lower two petals sometimes grown together. When an insect lands on the flower to take nectar, all the petals except the standard bend under its weight. This opens the keel and exposes the stamens and pistil, which touch the underside of the insect's body. The stigma (top of the pistil) projects and touches the insect first, receiving pollen which the insect has picked up from another flower. Then the stamens touch the insect, and transfer more pollen to the abdomen. When the insect flies off, the flower closes up again.

Bumblebees tend to fly from one flower to another of the same species, so cross-pollination is quite likely. Another advantage of this type of flower is that the stamens and pistil are protected from the rain.

The flower of the bird's foot trefoil has five petals: the standard is large and conspicuous, there are two wings at the side, and the keel, which is practically hidden is made up of two petals joined together. Inside the keel are the reproductive structures.

The ten stamens are first to ripen, and pollen falls into the keel. Below this mass of pollen is the stigma. When an insect lands on a flower, the keel is bent downwards by the weight, so that the pistil moves forward, pushing the pollen out so that it sticks to the abdomen of the insect. Only when the stamens have released all their pollen does the stigma itself ripen enough to receive pollen from in-coming insects.

The flowers of the tufted vetch are similar. The pollen lies in a trough-shaped narrow part of the keel. When the insect lands on the flower, the keel is pushed down, and the stigma comes forward. But in addition, there is a movement of the stamens behind the stigma. The stigma touches the insect first, and only after this is the pollen 'pushed' on to the insect's body.

7. Bird's-foot trefoil *Lotus corniculatus* **with a common blue butterfly** *Lycaena icarus*

This fairly common plant has a cluster of four to six golden yellow flowers in each inflorescence, which often turn reddish later. The flowers are pollinated in much the same way as clover flowers, and contain sticky pollen and nectar hidden inside. Insects with long probosces, mainly bees and butterflies, visit the flowers.

8. Tufted vetch *Vicia cracca* **with patchwork leaf-cutter bee** *Meganchile centucularis*

This vetch grows on grassland and roadsides. The inflorescence contains flowers about 1 cm long pollinated by butterflies and bees with long probosces. The patchwork leaf-cutter bee gathers edible pollen, using the 'brush' of stiff reddish brown hairs on its abdomen. See page 49, and the next column on this page.

9

10

UMBELLIFERAE

Both plants on this page belong to the Umbelliferae family. The flowers are small, but they grow in flat inflorescences of dozens or even hundreds of flowers. The whole inflorescence is conspicuous, and the nectar is available in large quantities near the surface of the flower.

Flies, wasps, bees, butterflies, and even beetles and ants walk about on the flowers, especially in hot weather. Much pollination takes place, including self-pollination, but each breath of wind blows away some of the visitors, which land on other plants.

People who study insects (*entomologists*) find Umbelliferae among the best places to catch specimens. As well as the wild carrot and wild parsnip, this family includes the parsleys, water parsnip, angelica and hemlock.

9. Wild carrot *Daucus corota* **with European leaf beetle** *Cryptocephalus sericeus*

Carrots grow in the wild as well as in cultivation. See also page 43.

10. Wild parsnip *Pastinaca sativa* **with** *Echinomyia grossa*

The wild parsnip grows in fields, verges and dykes. See also page 43.

11

11. Burnet rose *Rosa spinosissima* with garden chafer *Phyllopertha horticold*

The burnet rose, like most other roses, attracts insects by its scent, colour and shape. Scarcely any nectar is produced, but the many stamens produce a large quantity of pollen. The open flower offers an easy landing place for bees, flying beetles and flies. As well as walking about on the flower, these insects move into the centre. The stigmas are situated here, so they have a good chance of pollination. The flask-shaped receptacle then swells and grows into a rosehip.

The garden chafer is not a harmless visitor. As well as the leaves, it eats the petals, with the result that the flower withers. The larvae of this beetle are also harmful for they live underground and eat the roots. Since this stage lasts for two years, they can do quite a lot of damage. See also page 42.

12. Poppy *Papaver rhoeas* **with small garden bumblebee** *Bombus hortorum*

Poppies have no nectar and no scent, but they produce a large quantity of pollen from their many stamens. The stigmas are really the stripes on the top of the cylindrical or barrel-shaped ovary. The plant is sterile to its own pollen. Bees and bumblebees are common visitors, with occasional flies and beetles.

13. Lime *Tilia cordata* **with honey bee** *Apis mellifica*

The sweet smell attracts bees, bumblebees, wasps and flies, and the nectaries are easy to reach. The nectar is protected against the rain by dome-shaped sepals and the large bract on the stalk of the inflorescence. The stamens ripen first, and the stigma later, making cross-pollination more likely. See also page 43.

14. Ivy *Hedera helix* **with common wasp** *Vespa vulgaris*

Although the ivy does not flower until September, its inconspicuous flowers still attract insects. This is because they produce much nectar. The stigmas ripen after the stamens, and the plant is sterile to its own pollen. All kinds of flies, bees and wasps visit the flowers, because the nectar is easy to reach.

15. Flowering rush *Butomus umbellatus* **with hornet** *Vespa crabro*

The flowers have half-concealed nectaries at the foot of the pistils. The stamens ripen in two groups, first the outer six, then the inner three, and only after that do the stigmas ripen. Flies, bumblebees, bees and especially hornets, visit the flowers of this marsh plant throughout the summer. If all else fails, self-pollination is possible.

16. Swede *Brassica napus* **with honey bee** *Apis mellifica*

The flower has four long and two short stamens. The nectar collects between the ovary and the foot of the stamens, so that the liquid is fairly easy to reach. The stigma ripens before the stamens, but in any case the stamens turn their anthers away from the stigma before bursting open. Self-pollination only rarely occurs. Visitors are bees, flies and butterflies. See also page 44.

17. Yellow flag *Iris pseudacorus* **with a hoverfly** *Rhingia campestris*

The 'perianth' consists of 3 broad outer petals with a conspicuous 'honey guide', and three smaller upright ones alternating with them. Opposite each of the three broad petals is a long yellow curving branch of the stigma, and hidden between the two is a single large stiff stamen. A small sticky lip on the outside of each stigma lobe is the true stigma.

POLLINATION OF THE YELLOW FLAG

The three outer petals are broad; they attract the insects, and give them a good landing place. The pattern leads the insect to the centre of the flower where the nectar is produced. In order to get there, it has to push under a stigma-lobe; in doing so it presses against the little lip, so that the sticky side touches its back. If there is any pollen on its back, pollination takes place.

Even then the insect has not reached the nectar, so it presses on, under the stiff stamen. Pollen brushes off on to the hairy back of the insect. If the insect's proboscis is long enough, it can eventually reach the nectar.

As the insect retreats, the lip of the stigma is pressed against the stigma-lobe by the insect's back, with the sensitive side away from the insect, so the pollen which it has just picked up does not touch the stigma of the same flower.

It is also remarkable that not all the flowers of the yellow flag are pollinated by the same species of insect. Some admit large bumble-bees, while others can only admit slender hoverflies, such as *Rhingia campestris*.

18

18. Buddleia *Buddleia davidii* **with butterflies**

The buddleia came originally from China, but is now a popular garden plant. The tall purple inflorescences are most attractive. But some people plant buddleias because they attract butterflies. The brimstone butterfly (*Gonepteryx rhamni*, shown near the top of the inflorescence) can be found all the year round, even in winter. These butterflies survive the winter in hiding, and can be seen flying around again in the spring. This is also the case with the peacock butterfly (*Nymphalis io*, shown near the bottom of the inflorescence) and with the small tortoiseshell.

The red admiral (*Vanessa atalanta*, in the middle of the picture) and the painted lady (*Vanessa cardui*, behind the inflorescence on the left) come from warmer climates. They arrive as migrants in May or June, and a new generation is born here but does not survive the winter.

When butterflies suck nectar, we can clearly see their thin, curved, tube-like proboscis. It coils up in a spiral, and can be uncoiled when needed. Because grains of pollen sometimes stick to it, the butterfly often cleans its proboscis with other mouth parts after use.

The flower of the buddleia is shallow, and other insects besides butterflies are able to pollinate it.

19. Forget-me-not *Myosotis arvensis* **with small copper butterfly** *Lycaena phlaeas*

The heart of the flower is marked by five yellow scales, below which the stamens are hidden. There is a ring-shaped nectary on the receptacle which only insects with long probosces can reach. See also pages 27 and 45.

20. Viper's bugloss *Echium vulgare* **with wool-carder bee** *Anthidium manicatum*

This plant grows in dry and sunny places. The style and stamens project far outside the flower, and cannot help coming in contact with insect visitors. The nectar is deep inside the flower, but the funnel is wide. Bees are frequent visitors. See also page 27.

21. Ling *Calluna vulgaris* **with honeybee** *Apis mellifica*

Bees, wasps and flies enter the hanging flower from below. There are five dark coloured nectaries on the receptacle, which can only be reached by some pushing and rubbing, during which the pollen falls on to the visitor. It will be rubbed off on another visit. Wind pollination is also possible. See also page 45.

COLOUR CHANGE OF THE COROLLA

Unopened flowers are usually inconspicuous—often green in colour and not easy for an insect to see. This means that the insect wastes no time looking for nectar in unopened flowers. In a rather similar way, the flowers which have already been visited tend to get darker and duller, attracting fewer insects.

In the case of the forget-me-not and viper's bugloss, the corolla can be three quite distinct colours. The unopened flower is reddish purple, the open flower is a conspicuous blue or violet, and the ageing flower is a dull purple.

HILL CUCKOO BEES AS PARASITES

It is not just by chance that the hill cuckoo bee (*Psithyrus rupestris,* seen in picture 22) looks so much like the red-tailed bumblebee (picture 3). Red-tailed bumblebees live in underground nests, preferably made in heaps of stones.

The queen hill cuckoo bee does not herself make a nest, but invades a red-tailed bumblebee's nest. There she deposits eggs in some of the cells. The larva of the hill cuckoo bee first eats the bumblebee larvae, then lives on the honey and pollen brought in by the bumblebee workers.

22. Bilberry *Vaccinium myrtillus* **with hill cuckoo bee**

The flowers, and later the fruits of the bilberry are hidden under the leaves. Bees and bumblebees buzz about among the bushes, searching for the hanging, bell-shaped flowers which are rich in nectar. As the visitor reaches the flower from below, pollen drops on him. Self-pollination and cross-pollination are equally possible, and hybrids too can result.

23. Rosebay willow herb *Epilobium angustifolium* **with green hairstreak** *Callophrys rubi*

Willow herb flowers grow in tall inflorescences. The stamens of each flower ripen first, sticking horizontally beyond the 'corolla', while the style with its closed stigma is still pointed down. Later the stamens bend down and the stigma lifts itself up to take their place. Many insects visit the plant, causing cross-pollination. See also page 48.

24

24. Purple loosestrife *Lythrum salicaria* **with swallowtail butterfly** *Papilio machaon Britannicus*

This marsh plant grows in large clumps in rush vegetation, using runners to reproduce itself. Pollination between flowers within the clump would be a kind of self-pollination, because the plants of the group all grow from the same root system. But in fact the purple loosestrife has three different kinds of flower, and successful fertilisation can only occur between two flowers of different types.

Flowers of type A have a long style, six medium-length stamens and six short stamens. Type B has six long stamens, a medium-length style and six short stamens. Type C has six long stamens, six medium-length stamens and a short style. See page 47 for diagrams of the three types of flower.

Pollen could fall from the stamens on to the stigma in types A and B, but it is self-sterile and no fertilisation occurs.

The swallowtail butterfly is now quite rare, and in Britain is only found in the fens and broadland of East Anglia.

25. Cowslip *Primula veris* **with flower bee** *Anthophora acervorum*

Only bumblebees, bees and butterflies with fairly long prosbosces can reach the nectar in these flowers. There are two types of flowers, which never occur on the same plant, and fertilisation is only successful when pollen from one type reaches the other. See also page 47.

26. Grape hyacinth *Muscari racemosum* **with honey-bee**

The picture shows an experiment to find whether a worker honeybee follows the scent of the flower, or whether it relies on its eyesight at close range. The glass tube is open below, but the bee keeps flying against the glass.

27. Figwort *Scrophularia nodosa* **with field digger wasp** *Mellinus arvensis*

The pistil of the cup-shaped flowers ripens first, while the four stamens are still curled up inside. Later the pistil droops a little and the stamens rise. Flowers with brownish colours are sometimes visited by wasps and by digger wasps such as the one illustrated. See also page 48.

28. Grass of Parnassus
Parnassia palustris **with currant hoverfly** *Syrphus ribesii*

This plant grows in wet moorland and in sand dunes. The newly opened flowers have five petals, white and green-veined, round a central pistil. Five stamens also surround the pistil. In addition, at the base of each petal, there is a large greenish nectary with a number of branches, each carrying a sticky yellow droplet.

29. Grass of Parnassus with swarming hoverfly *Scaeva pyrastri*

In favourable weather one stamen will ripen each day for five days. The anther of the ripe stamen bursts open. After some hours the ripe stamen bends away from the pistil, and its empty anther then falls off. Each day one more anther is discarded, and in this way one can tell the age of the flower. After five days there are five beheaded stamens.

30. Grass of Parnassus with hoverfly *Syrphus balteatus*

The nectar is produced at the base of the nectaries, but flies are first attracted by the shiny yellow droplets on the branches. These provide no food, but if the flies walk about on the flower to find the nectar, they cause pollination. See also page 46.

31. Wild pansy *Viola tricolor* with Queen of Spain fritillary *Argynnis lathonia*

The flower construction of the violets and the closely related pansies is suitable for pollination only by butterflies and a few species of bee, because a long proboscis is needed to reach the nectar. The flower has an easily recognisable shape (to us it looks rather like a face) and the centre of the flower is clearly marked.

If we examine such a flower from the outside, we find five green sepals which make up the calyx and five green petals, the lowest one of which often has honey guides on it and extends behind the flower into a spur where the nectar collects. In the centre of the flower is a rounded ovary, and surrounding it are five short stamens, two of which have hair-like nectaries extending into the spur. Projecting forward from the ovary is a short style, and at its tip a bud-like stigma. This stigma has a tiny pit with a sticky flap beneath it. Pollen is shed from the anthers and collects in the narrow opening between the stigma and the nectar in the spur.

A visiting insect first touches the sticky flap on the stigma, and then collects the pollen on its way to the nectar. As it withdraws it closes the stigma flap over the tiny pit and so cross-pollination is made possible, and self-pollination is prevented.

In spite of their vivid colours, violets and pansies attract remarkably few visitors. It seems as though they may have over-specialised in insects with long probosces.

Surprisingly, some violets and pansies can pollinate themselves in another way, for they have a number of flowers which never open at all, but which still produce seed by self-fertilisation.

32. Red campion *Melandrium rubrum* humming-bird hawk moth *Macroglossa stellatarum*

The flowers of the red campion open towards dusk, and are still open the next morning, but then they soon close or wither. The five pink or white petals are supported at the base by a firm bulb-shaped calyx, and form a suitable landing place for moths. On any one plant all the flowers are either male or female. Self-pollination is impossible.

Because the nectar is produced at the bottom of the fairly deep flower, it can only be reached by insects with long probosces, or by insects with medium-length probosces but strong enough to push deep into the flower. These are mainly certain species of moth.

MOTHS

If you sit in a field after dark with a lamp, you will find that there are many moths fluttering about. They are attracted to the lamp because they instinctively fly towards a light, especially one giving out ultraviolet light. But lamps are man-made; in the dark, moths react to the faint gleams of light reflected from flowers, and they also respond to scents.

Night-flying moths include the humming-bird hawk moth (32), the poplar hawk moth *Laothoe populi* (33) and the lime hawk moth *Mimas tiliae* (34). These large moths fly vigorously round a lamp, but they can hover over a flower, fluttering their wings fast like a humming bird while their long proboscis sucks out the nectar.

THE NOTTINGHAM CATCHFLY

Found mainly in coastal areas, the catchfly has flowers with both pistils and stamens, but some of the plants have only male or female flowers. As the nectar lies deep down, and the fragrant white flowers are open only at night, they are almost all pollinated by moths. Only five of the ten stamens of the bisexual flowers are ripe on the first day of flowering, five on the second, and the three stigmas on the third day. When the flowers age, they take on the 'nodding' day position.

THE WILD HONEYSUCKLE

The flower opens up late in the afternoon, while there are still bees and bumblebees about. Its scent is very strong at first, and the narrow funnel of the flower fills up with nectar. The first visitors are therefore always bees or bumblebees, which can just reach the nectar. When these insects have returned to their hives, the moths have their turn. Their long prosboces can reach to the bottom of the funnel. In the morning the day insects return to see if anything is left. As the stigma and stamens do not ripen at the same time, and the stigma protrudes beyond the stamens, there are good chances of cross-pollination. See also page 50.

33. Nottingham catchfly *Silene nutans* **with** *Laothoe populi*

n daytime the flowers are pendulous and shrunken. About p.m. they open out, and close gain about 8 a.m.

34. Honeysuckle *Lonicera periclymenum* **with lime hawk moth** *Mimas tiliae*

The scent and nectar attract bumblebees by day and moths by night.

35

36

35 & 36. Meadow sage or clary *Salvia pratensis* with bumblebee

The blue flowered meadow sage, or meadow clary, is becoming quite rare now, but garden sages (also with blue flowers) have the same flower structure. The flower has a short tube with two lips—a curved upper lip and a thickened lower lip. The lower lip is a good landing place for bees and bumblebees, which are attracted to blue flowers and can reach the nectar fairly deep inside the flower. The upper lip protects the stamens and nectaries against rain.

Each flower has two stages. When it is young, the style is short, and the stigma not yet sticky, but the anthers are already open and pollen can stick to a visitor. Later the style protrudes further, and the stigma is now sticky. This helps to bring about cross-pollination. See also pages 35 and 50.

37. Ground ivy *Glechoma hederacea* with red osmia *Osmia rufa*

This is another blue-flowered plant with similar lips, common in fields and along road-sides. Some of the plants are bisexual and some are just female. The bisexual flower is clearly marked by two-lobed anthers, one above the other. All have the nectar hidden fairly deep inside the flower tube. The visitors are bees and bumblebees. See page 50.

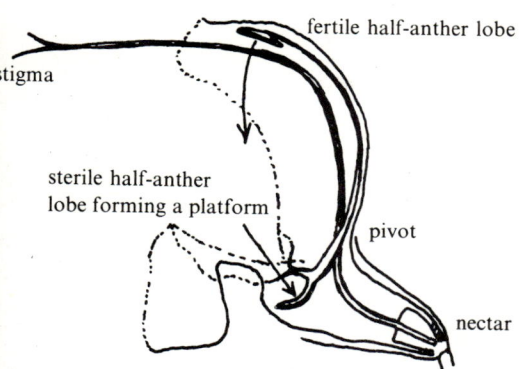

Section through the flower

THE 'POLLEN TIPPER' OF THE MEADOW SAGE

The meadow sage has two stamens, each with a short filament and two anther lobes of unequal size. The two short anther lobes contain no pollen but form a platform at the mouth of the corolla tube. The other two anther lobes contain pollen and are borne on slender stalks beneath the upper lip. Each half-anther is pivoted on the short filament. A visiting insect lands on the lower lip of the corolla; as it pushes into the flower to reach the nectar, it presses on the shorter anther lobes, bringing down the fertile lobes on to its back, with a tipping movement like a see-saw. The pollen is then at exactly the right place to reach the stigma when the insect visits an older flower. The stamens shed their pollen before the stigmas are ripe.

38. Columbine *Aquilegia vulgaris* with red-tailed bumblebee *Bombus lapidarius*

This garden plant is found in many colours. The drooping flowers have five sepals and five petals, which alternate with each other and are all conspicuously coloured. Each petal has a long nectar spur, curving inwards towards the stalk. Only insects with very long tongues can reach the sweet nectar.

39. Larkspur *Delphinium aiacis* **with bumblebee**

The five bright 'petals' of the delphinium flower are in fact sepals. Together they form a deep nectar spur. In addition, there are two small petals, grown together, which form a second spur within the first. They are only visible when the flower is dissected. The shape, colour, and deeply hidden nectar, ensure that most visitors will be bees and bumblebees.

40. Monkshood *Aconitum anglicum* **with small garden bumblebee** *Bombus hortorum*

These flowers are pollinated almost solely by bumblebees. The flower has five conspicuously coloured sepals. The upper one is hood-shaped, and inside this there are two petals, which are deformed into long-stemmed, curved nectaries. The many stamens ripen first and the pistil later. See also page 48.

THE RANUNCULUS FAMILY

The flower forms of the plants in this family vary enormously. For a long time it was hard for botanists to believe that they were related to one another.

Firstly, there are the 'pollen flowers'. The anemones have no nectaries, but their many stamens produce so much pollen that a vase of them is very soon surrounded by a layer of pollen. The wood anemone, *Anemone nemorosa,* the meadow rue. *Thalictrum* species, and old man's beard, *Clematis vitalba* are visited only by pollen-gathering bees.

Then there are insect flowers with nectar near the surface. The marsh marigold *Caltha palustris* which flowers early and has little nectaries at the base of each petal is a good example. The many kinds of buttercup *Ranunculu. species* have frequent visits from flies and butterflies.

Some members of the family have deep-hidden nectar. There is the group of spurred flowers such as the delphinium (39) and the columbine (38) with a number of curved nectar-containing spurs. There are those with nectar containers which are actually specially shaped petals, like the monkshood.

The lesser celandine is a special case. It has abundant flowers but is scarcely ever successfully pollinated. It reproduces by means of 'bulbils', which are small tubers appearing either from the leaf axils or from the roots. To ensure survival, they no longer need to flower at all.

POLLINATION OF ORCHIDS

The 'pillar' in the centre of an orchid flower is in fact the style with its two stigmas joined to the single anther with its two anther lobes. Each visiting insect bumps into the central pillar. At the base of the front of the pillar are two short stigma-lobes. Higher up there is the single two-lobed anther. In each lobe, the pollen sticks together to form a little lump on a stalk, which ends in a sticky disc at the bottom. These are called 'pollinia'.

The insect visitor pushes under the pillar in search of nectar, and the stigma receives any pollen which is already on its head. As it withdraws, it bumps against the base of the stamen and the sticky pollen clumps are withdrawn and may stick to its head. Sometimes insects become almost covered in pollinia; they are correctly placed to pollinate the next flower.

41. Early purple orchid *Orchis mascula* **with rose chafer** *Cetonia aurata*

This has spotted leaves, and flowers in April-June. What looks like a flower stalk is in fact an inferior ovary. On this, placed centrally within the flower, is the thick pillar which contains the pistil and stamen. Each flower has a nectar spur. See also page 51.

42. Marsh helleborine *Epipactis palustris* **with common wasp** *Vespa vulgaris*

Like the early purple orchid, this occurs in marshy grassland, amongst rushes and in dunes. Neither is common and specimens should not be picked. The lowest petal has a broad projecting ridge, behind which nectar collects. Wasps carry out the complicated pollination of these flowers. See also page 51.

43

43. Birthwort *Aristolochia clematitis* with harlequin fly *Chironomus plumosus*

This is a rare and localised plant, but its unusual method of pollination makes it worth inclusion here. The flower is a long tube, with a sloping funnel at the outer end, making a good landing place for visiting gnats and small flies. They are attracted by a faint smell, crawl down through the tube and arrive inside a ball-shaped hollow. In this there is a knob, formed by a number of stamens grown together with a stigma. The stigma ripens first. If the insect is carrying pollen, the stigma will be pollinated.

The insect came in search of nectar, but there is none there, so it tries to crawl back through the tube.

But stiff hairs, pointing inwards, make this impossible and it is trapped there for several days! Meanwhile the anthers ripen, and as the insect makes desperate attempts to escape it is covered in pollen. Then the hairs which prevent it getting out become limp. It can crawl out, but may well fall into a similar trap again. If it enters another birthwort flower, it will pollinate it and may even die inside it. See also pages 46–7.

44. Lords-and-ladies *Arum maculatum* **with owl midge** *Psychoda alternata*

The lords-and-ladies (also called cuckoo-pint and wild arum) is common in British hedgerows, except in Scotland where it is localised. Its method of pollination can also be observed in garden varieties of arum lily.

What looks like a flower is, in fact, a large bract which is hood-shaped, but has a swollen base surrounding a central column or 'spadix'. This spadix is fleshy with very tiny stalkless flowers on the part inside the base and a club-shaped top protruding above. From the bottom of the spadix we find, firstly female flowers with pistils and a few hairs, then male flowers with stamens, then another ring of stiff hairs, facing down, and finally a stalk ending in the club-shaped swelling.

When the bract has unfurled, it acts as a lure for small flying insects. They are also attracted by a faint smell given off by the spadix during flowering. When the insects fly under the hood, they find that the walls are very slippery, and they slide right down into the base below. They crawl around the stigmas, which are ripe for pollination. It is no use trying to climb out, because the ring of hairs effectively stops this. When the stamens ripen, and shed their pollen, the walls become less slippery and the hairs become limp, allowing the insects to escape.

The insect may have been imprisoned for hours or even days, and at last it can climb out. In doing so, it will crawl over the ripe anthers just under the ring of hairs, and will collect pollen. Self-pollination is quite impossible. Later in the year the plant produces many red berry-like fruits. See also pages 46–7.

45. Spotted deadnettle *Lamium maculatum* with long-horned eucera *Eucera longicornis*

Deadnettles are extremely versatile in their methods of pollination. If one method fails, another may succeed. The flowers—white, purple or spotted—are tube-shaped at the base. At the foot of this tube are hidden nectaries, well protected by the hood-shaped upper lip and by a dense crown of fine hairs. The scent, colour and striking shape of the flowers all attract visitors.

Insects find the flowers easy to see because they are in clearly visible groups, the lower lip protrudes to give an easy landing place, and the arrangement of leaves gives a clear flight-path to the flowers.

Usually the lower lip of the flower has a conspicuously coloured honey-guide, showing the way to the nectar. The style sticks out beyond the stamens, promoting cross-pollination. The visitor first comes into contact with the stigma, and afterwards with the anthers, which brush his back.

As the insect retreats, it carries pollen away on its back, passing the stigma, but at worst it will touch the underside of the stigma which is not sticky.

The visitors are confined to certain bumblebees and bees, all of which tend to visit the same kind of flower over a long period, so the depositing of pollen from other species is not very likely.

If the nettle is not pollinated by insects, the style begins to bend, and the stigma comes into contact with the flower's own pollen. In this way, self-pollination is possible.

45

Classification

Even if we consider only insect-pollinated flowers growing in Britain, the number of species is vast. We can classify them into groups of broadly similar flowers but in nature no sharp boundaries can be drawn. There are always plants which do not quite fit into one category or another.

1. Pollen flowers. No nectar is produced but there is much edible pollen (e.g. roses, poppies).

2. Flowers with nectar easily available to all insects. Some of these plants have individual flowers and some have flowers in dense inflorescences (e.g. lime, wild carrot and wild parsnip, ivy, willow).

3. Flowers with nectar half-hidden (e.g. swede and turnip).

4. Flowers with hidden nectar, available only to insects with long tongues. Some are individual flowers; others grow in inflorescences of varied size. (E.g. violets, buddleia, forget-me-not, viper's bugloss, ling, colts-foot and cornflower).

5. Flowers adapted for pollination by flies and gnats (e.g. grass of Parnassus, birthwort, and lords-and-ladies).

6. Flowers adapted for pollination by particular groups of bees and bumblebees (e.g. clover, vetch, bird's-foot trefoil, sage, ground ivy and deadnettle).

7. Flowers adapted for pollination by butterflies and moths (e.g. red campion, Nottingham catchfly and honeysuckle).

8. Other categories not included here are flowers pollinated by wind and water, by birds, by snails and by bats.

Tongue length of visitors

To give some idea of the depth various insects can reach, here is a list of the tongue lengths of various insects. The numbers in brackets refer to the pictures on pages 17–40

Sun fly (1), currant hoverfly (28), swarming hoverfly (29), *Syrphus balteatus* (30), *Echynomia grossa* (10), are all between 2–5 mm. Slender hoverfly (17) reaches 17 mm.

The mining bee (4), leaf-cutter bee (8), honey bee (13, 16, 21 and 26) and red osmia (37) are between 5–8 mm. The wool-carder bee (20) reaches 10 mm and the flower bee (25) as much as 20 mm. The red-tailed bumblebee (3 and 38) is between 10–14 mm, and small garden bumblebee (40) 15–20 mm.

Butterflies such as the red admiral (18), small tortoiseshell (2), peacock (18), painted lady (18) and brimstone (18) have tongues which stretch to 14–17 mm. The humming bird hawkmoth (32) reaches 26 mm and the lime hawk moth (34) 30 mm.

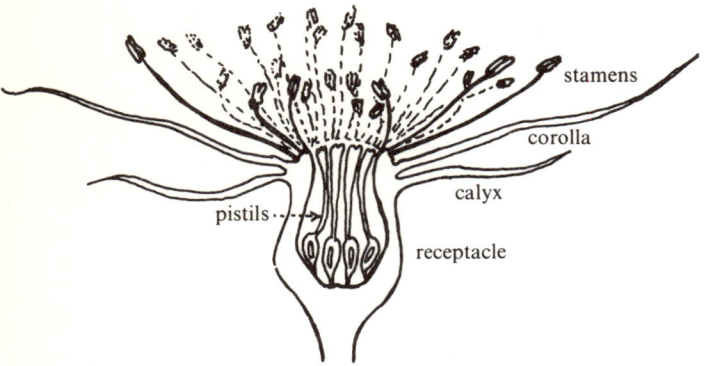

Cross-section through the open flower of a rose

Roses (see page 22)

On the left is a diagrammatic cross-section through a rose. What finally becomes the rose-hip is a fleshy receptacle with a number of single-seeded ovaries inside it. These later become the true fruits. The hip is only a 'false fruit'.

No nectar forms inside the flower, but there is an excessive amount of pollen which is collected by bees. As the visitor crawls about, it comes into contact with the stigmas. A great deal of pollen is used as food for the visitors, but there is enough left for pollination. The fleshy rosehips which form later are eaten by birds and the small fruits inside are dropped. This is called fruit-dispersal by birds.

Poppies (see page 23)

In the centre of the flower there is a pistil (drawn larger than life in the simplified drawing) which consists of a barrel-shaped ovary with a flat lid. The stripes on the lid are the stigmas, which are sticky when ripe. An insect bringing pollen may alight on these stigmas. Then it may collect more pollen from the stamens.

The ovary later grows into a 'pepperpot' type of fruit distributing much fine seed from the holes as it is blown about by the wind.

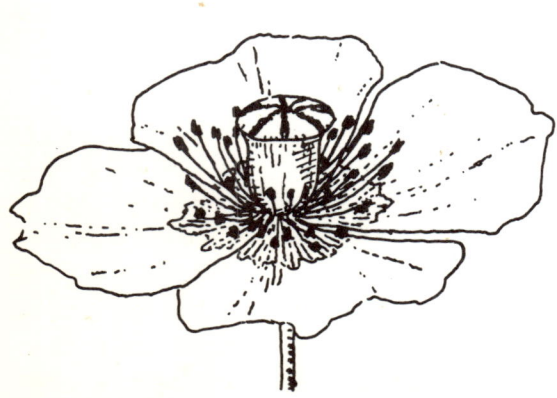

Simplified drawing of a poppy flower

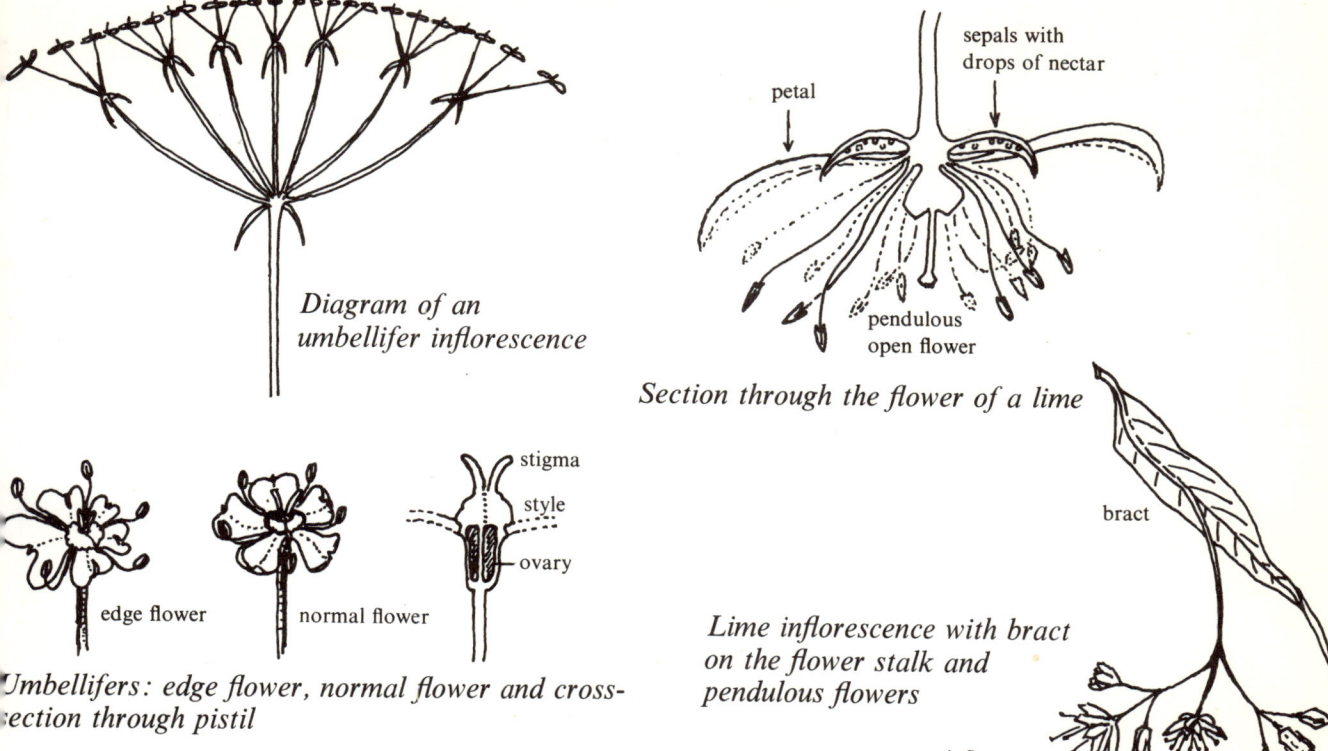

Diagram of an umbellifer inflorescence

Section through the flower of a lime

petal

sepals with drops of nectar

pendulous open flower

bract

stigma
style
ovary

edge flower

normal flower

Umbellifers: edge flower, normal flower and cross-section through pistil

Lime inflorescence with bract on the flower stalk and pendulous flowers

inflorescence

Umbellifers (see page 21)

These include wild parsnip and wild carrot, and come into category 2 on page 41. Their flowers are usually in a dense cluster, but together form a flat 'table-top'. Insects are attracted by the large number of flowers with nectar near the surface. The flowers on the outside have one petal larger than normal, to make them easier to see. In the section through the pistil, we can see how the two stigmas thicken, towards the base, into nectaries.

Lime (see page 23)

This also belongs to category 2. Its flowers are fairly far apart, they hang downwards, and insects have to be able to fly in from underneath, and cling to the centrally placed stigma. They may pollinate it and, in reaching up for the nectar, they may touch the stamens. Bees, which visit just one type of flower at a time, are particularly attracted to the lime flowers.

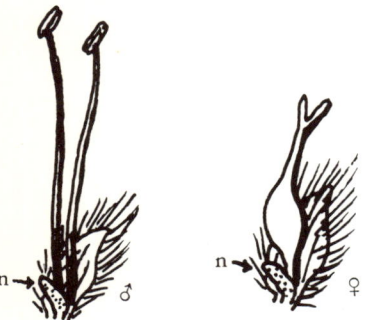

Male flower and female of the willow. Nectary is shown by the letter 'n'

♂ catkin

♀ catkin

Catkins of male willow and of female willow

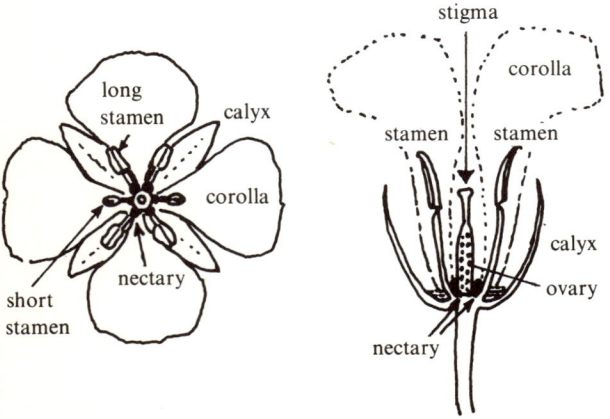

Left: diagrammatic plan of a crucifer. Right: section through a crucifer

Willow (see page 18)
Willow flowers have no corolla and no calyx. They are simply the male or female organs, supplied with a nectary. Each flower stands in a bract, which protects it against rain. The nectar is fairly near the surface, and willows come into category 3.

Swede (see page 24)
Most species in the crucifer family (which includes the cresses, mustard and radish) have fairly shallow nectaries placed at the foot of the six stamens (four long and two short). There is more nectar in sunny weather, and it often collects at the hollow foot of the four sepals. The shape, colour and scent are easily recognised by many insects, especially the honeybee. Category 2.

Coltsfoot (see page 17)
The illustrations at the top of the next page show a female flower from the outer edge of the inflorescence; a bisexual flower from the centre and a section through the head. Many flowers in the same family have strap-shaped ray florets and tube-shaped disc florets. The tube florets offer nectar, but the visitor needs a long tongue, so they fall into category 4.

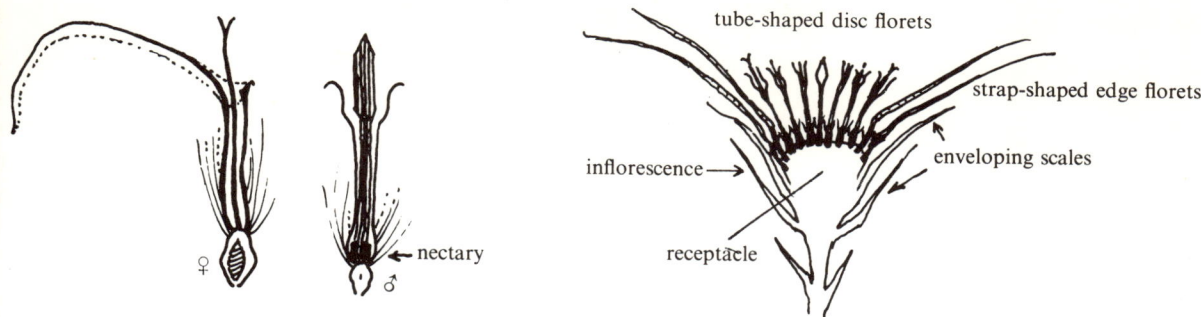

Left: ray floret of a coltsfoot. Right: disc floret of a coltsfoot

Section through the inflorescence of a coltsfoot

Ling (see page 26)

The ling also comes into category 4, but the nectar is harder to reach than in the coltsfoot. The insect must fly up and in from the side, and hold on firmly. The visitor is sprinkled with pollen on its back, whereas the visitor to the coltsfoot 'walks' through pollen and picks it up on its legs.

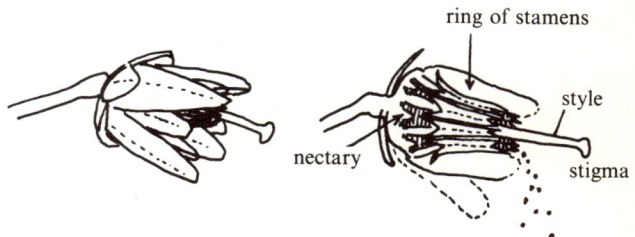

Pollen distribution in the ling

Forget-me-not (see page 26)

In the various kinds of forget-me-not, the nectar is produced at the foot of a corolla which has grown into a tube. A fairly long tongue is needed to reach it. It is made especially difficult to reach because there are scales on the edge of the corolla tube, which project inwards. But these scales are yellow, in contrast with the blue corolla, and this may help in identification.

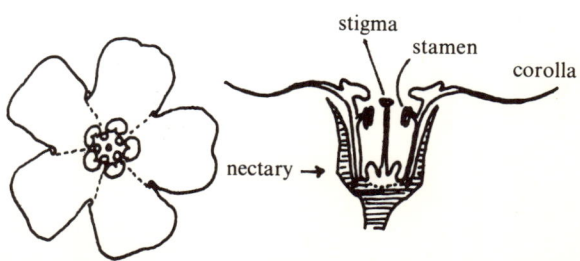

Plan view and section through a forget-me-not

Birthwort and lords-and-ladies (see pages 38–39)

The method of pollination of these plants was fully described on pages 38–39. In warmer regions much larger species of both kinds occur. Sometimes their hollow flowers teem with insects. The temperature inside the swollen base can rise to 42°C, and the larger relatives of the lords-and-ladies have a strong and unpleasant smell.

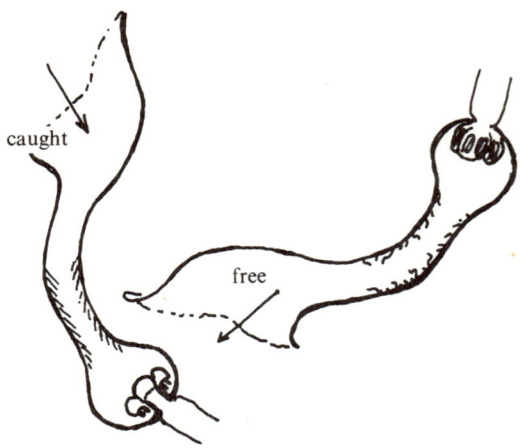

Sliding fall and hair collar of the birthwort

Grass of Parnassus (see page 30)

This plant is found in damp moorlands and is visited by flies. It therefore belongs to category 5.

The stumpy pistil in the centre of the flower has a four-lobed stigma and either no style or a very short one. Round it are five true stamens and five false stamens or 'staminodes' alternating with them. These have become the nectaries. The nectar is secreted at the base of each staminode but the rest of the nectary has developed fine hair-like branches, each with a tiny shining yellow drop at the tip. This is not nectar but it helps to attract insects to the flower.

The stamens do not all ripen at the same time, and in fine weather they shed their pollen very quickly. Picture (28) on page 30 shows a newly opened flower with five unripe stamens.

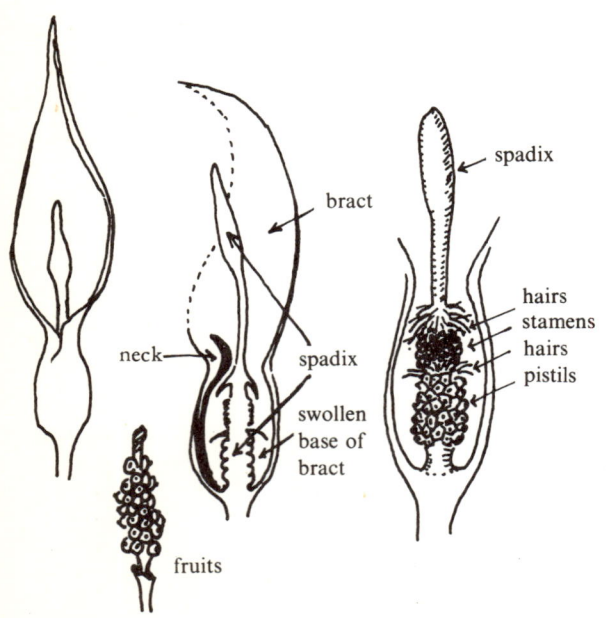

Lords-and-ladies

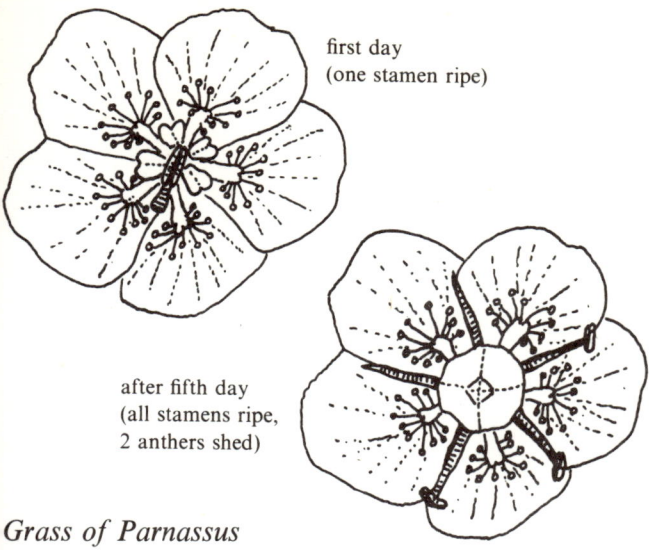

first day
(one stamen ripe)

after fifth day
(all stamens ripe,
2 anthers shed)

Grass of Parnassus

Picture (29) shows a flower with four ripe stamens and one still not ripe. Once the pollen has been shed, the anther falls away from the filament. On this page the diagrams show a flower with only one ripe stamen and another in which two anthers have already lost their pollen and been discarded.

Cowslip (see page 29)

There are several kinds of wild primula (cowslip or primrose) as well as hybrids. All have two kinds of flower, and belong either in category 6 or 7, as they are visited by both bees and butterflies.

Purple loosestrife (see page 28)

This too is on the borders of categories 6 and 7. The nectar is too deep for most insects other than bees, bumblebees and butterflies.

The three kinds of flower vary in the length and shape of the pistil, and also in the size of the pollen grains. If an insect visits a flower of type A, pollen from the long stamens may have stuck to his tongue. If he next visits flower B there is no pistil to pollinate at the same height on his tongue, but if he visits flower C the pistil will be at the right height to receive the pollen. The chance of cross-pollination occurring in this way is quite high.

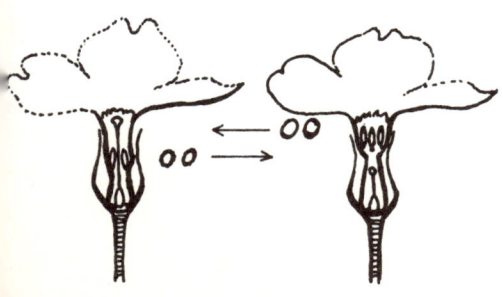

The two kinds of cowslip flower

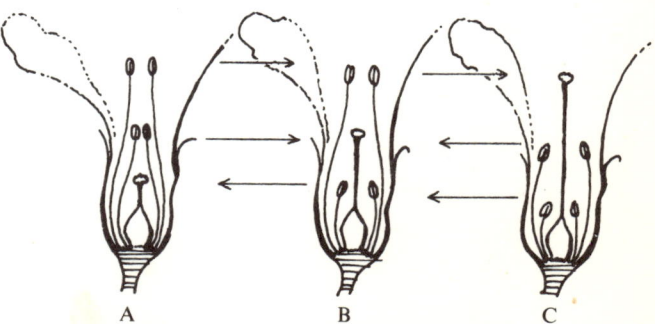

A   B   C

The three kinds of purple loosestrife flower

Willow-herb (see page 27)
This is on the border between category 2 and category 4. Both flies and butterflies make visits. To prevent self-pollination the stamens ripen first and the pistils later. This is called *protandry*.

Common figwort (see page 29)
This belongs in category 4, but could perhaps be placed in category 6 because it attracts wasps. Again, the stamens and pistils are not ripe at the same time, but here the stigmas are ripe before the stamens. This is called *protogyny.*

Monkshood (see page 36)
Both the garden monkshood and the wild yellow monkshood depend on the bumblebees alone. Apparently only the bumblebees can penetrate these flowers to reach the nectar, which is very high up in the two nectaries. The stamens ripen first, and, although there is no noticeable change in position, the stigmas stick out further once the stamens have withered.

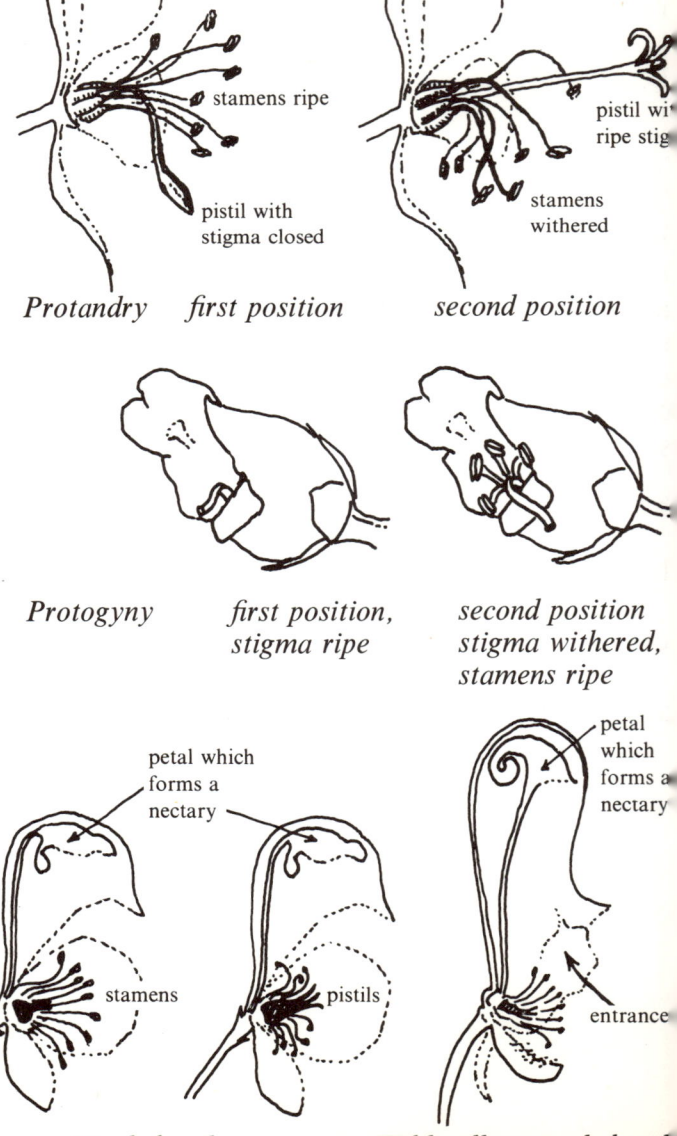

stamens ripe

pistil with ripe stig

pistil with stigma closed

stamens withered

Protandry   *first position*   *second position*

Protogyny   *first position, stigma ripe*   *second position stigma withered, stamens ripe*

petal which forms a nectary

petal which forms a nectary

stamens

pistils

entrance

Monkshood   *Wild yellow monkshood*

Papilionaceous flowers (see pages 19–20)
Most of the species in this family are typical insect flowers, mainly belonging to category 6 because they are visited by bees and bumble-bees. A few, such as bird's-foot trefoil, are also visited by butterflies.

The diagram on the right shows the five petals. The keel, formed of two petals, plays an important part in pollination. In flowers with no nectaries, the ten stamens have grown into a cylinder. In flowers with nectaries, nine stamens have grown into a little 'boat', while one stamen is free-standing. Lying within the cylinder or the boat is the pistil. There are several methods of pollination.

a. The mechanism described on page 19 and illustrated on the right.

b. The mechanism of the bird's-foot trefoil, described on page 20.

c. The mechanism of the tufted vetch, described on page 20.

d. The mechanism found in the broom. The flower is without nectar but is visited for pollen. The visitor presses the flower open. First five small stamens come shooting up, then the long stamens together with the style. This takes place quite forcefully, and only happens at the first visit.

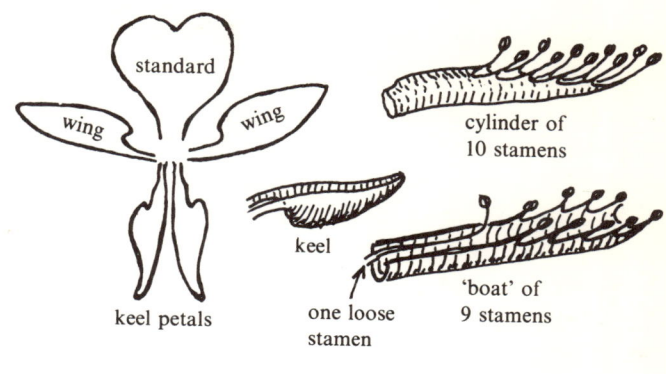

The mechanism of the red clover

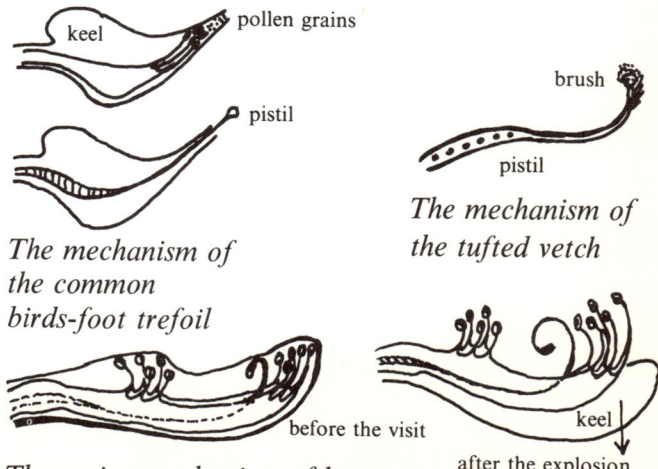

*The mechanism of
the common
birds-foot trefoil*

*The mechanism of
the tufted vetch*

The spring mechanism of broom

Labiate flowers (see page 34 and the drawing on page 8)

All members of this family have a tube-shaped corolla with an upper and a lower lip. Some have a short upper lip and a shallow corolla tube, for example, thyme. Other species have a larger upper lip, for example ground ivy, and are often visited by bees. Ground ivy has a 'tipping' system of the stamens (described on page 35). The anthers are always positioned to cover the back of an insect with pollen, and the pistil which ripens later projects far outside the flower so that it is the first part to touch the visitor.

Wild honeysuckle (see page 33)

The corolla tube is long and narrow. This shape is typical of the flowers of category 7. Honeysuckle is a 'night' flower, visited by moths. Hawkmoths especially, with their long tongues, hover just in front of the flower while sucking, but bees also visit the flower. They do so mostly when the tube is nearly filled with nectar, and long tongues are not needed.

The scent is strongest towards evening, and the newly opened flowers are very light in colour. Later they change to yellowish white. The fact that they are climbers makes their flowers easier to see.

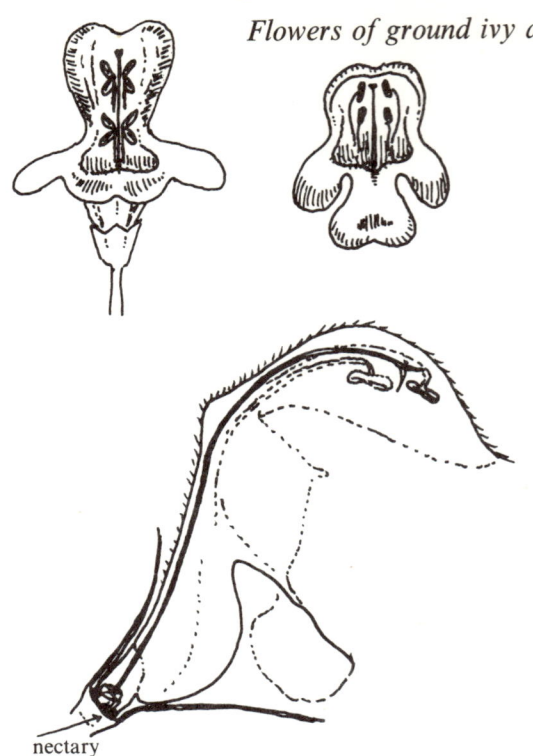

Flowers of ground ivy and thyme

nectary

Section through deadnettle flower

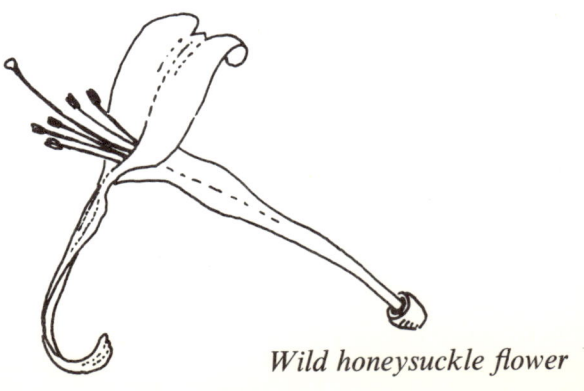

Wild honeysuckle flower

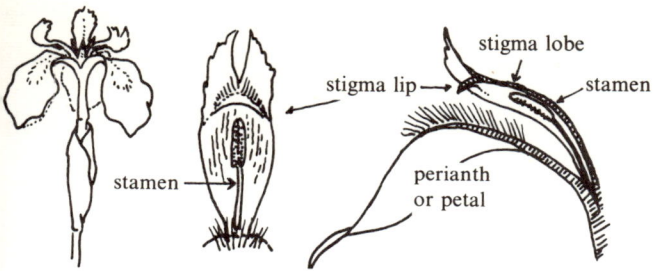

From left to right: side view of yellow flag; front view of a stigma-lobe with stigma lip and a stamen; section through stigma-lobe with perianth

Yellow flag (see page 24)

The flower has three landing places—the three largest and most strikingly coloured petals. The smaller petals stand upright and take no part in pollination. The three yellow petal-shaped stigma-lobes each have a stigma lip, sensitive to pollen.

Bumblebees are the insects which find visiting easiest, though long-tongued flies are also among the regular visitors.

Orchids (see page 37)

Compared with exotic greenhouse orchids, our wild species are not dramatic, but seen at close range with a magnifying glass they are very interesting. The forty to fifty species in this country vary a great deal in shape and colour. They also produce many hybrids, which makes them difficult to distinguish.

The method of pollination was described on page 37. Most orchids are pollinated by wasps, and belong to category 6. Those with a nectar spur can be pollinated only by an insect with a long tongue. The cultivated *Cypripedium* orchids have a pouch-like lower lip which serves as a slide; the victim can only escape by crossing the reproductive organs of the flower.

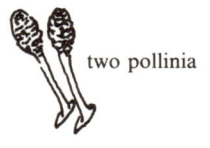
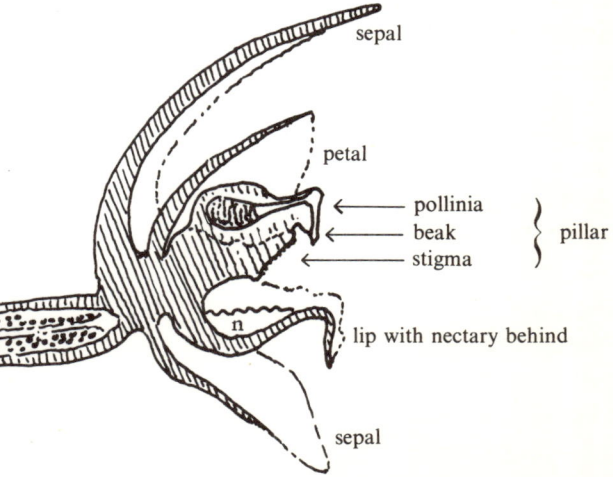

Left: flower of marsh helleborine. Right: pollinia and section through flower

Glossary

anther the part of the stamen containing the pollen grains

bisexual flower flowers with pistils and stamens

bract scale leaf

bulbils usually swollen tubers by means of which the plant can reproduce itself vegetatively

calyx the sepals as a whole

corolla the petals as a whole

fertilisation growth of the pollen tube and fusion with the ovule

filament part of stamen which carries the anther

honey nectar which has been regurgitated by the bee

inflorescence flowering branch

keel two lower petals in *Leguminosae* flowers

leaf axil place where leaf arises from the stem

nectar a sugary solution

nectary a gland which produces nectar

ovary part of pistil containing the ovules

ovule contains the 'egg' cell and develops into a seed after fertilisation

perianth floral leaves as a whole, including petals and sepals

petal a member of the inner whorl of perianth —usually brightly coloured

pistil female part of the flower

pollination transfer of pollen to stigma

pollinia contents of one anther lobe stuck together (as in orchids)

proboscis mouth parts of insect modified for sucking

protandry stamens ripe before stigma

protogyny stigma ripe before stamens

receptacle top of flower stalk—it may be flattened (coltsfoot) or cup-shaped (burnet rose)

sepal a member of the outer whorl of perianth —usually green

spadix found in lords-and-ladies, it carries all the pistils and stamens

stamen male part of the flower

staminode false stamen as in Grass of Parnassus

standard largest petal in *Leguminosae* flowers

stigma sticky surface usually on top of the pistil

style part of pistil between ovary and stigma

'tongue' (of insect) mouth parts modified to suck nectar

tubers swollen stems or roots

Other books about flowers and insects

Concise British Flora W. Keble Martin (Michael Joseph)

Pocket Encyclopaedia of Wild Flowers in Colour M. S. Christiansen (Blandford)

Field and Meadow Life L. Lyneborg (Blandford)

Finding Wild Flowers R. S. R. Fitter (Collins)

Why Flowers? Peter and Patricia Mann (Methuen)

Look at a Flower T. Dowden (Longman Young Books)

Insects J. Clegg (Muller)

Create a Butterfly Garden L. H. Newman (World's Work)

Name this Insect E. Fitch Daglish (Dent)

Insects in Colour N. D. Riley (Blandford)

Insects M. Prior (Black's Picture Information Books series)

Index of Latin names

All numbers in this index refer to pages, not to picture numbers.

Aconitum anglicum (monkshood) 36
Andrena haemorrhoa (mining bee) 18
Anemone nemorosa (wood anemone) 36
Anthidium manicatum (wool-carder bee) 26
Anthophora acervorum (flower bee) 29
Apis mellifica (honey bee) 23, 24, 26, 29
Aquilegia vulgaris (columbine) 35, 36
Argynnis lathonia (Queen of Spain fritillary) 31
Aristolochia clematitis (birthwort) 38
Arum maculatum (lords-and-ladies) 39

Bombus species (bumblebee) 34, 36
Bombus hortorum (small garden bumblebee) 23, 36
Bombus lapidarius (red-tailed bumblebee) 17, 18, 35
Bombus pratorum (early bumblebee) 19
Brassica napus (swede) 24
Buddleia davidii (buddleia) 25
Butomus umbellatus (flowering rush) 24

Callophrys rubi (green hairstreak) 27
Calluna vulgaris (ling) 26
Caltha palustris (marsh marigold) 36
Centaurea cyanus (cornflower) 17
Cetonia aurata (rose chafer) 37
Chironomus plumosus (harlequin fly) 38
Clematis vitalba (old man's beard) 36
Cryptocephalus sericeus (European leaf beetle) 21

Daucus carota (wild carrot) 21
Delphinium aiacis (larkspur) 36

Echinomyia grossa 21, 41
Echium vulgare (viper's bugloss) 26
Epilobium angustifolium (rosebay willow herb) 27

Epipactis palustris (marsh helleborine) 37
Eucera longicornis (long-horned eucera) 40

Glechoma hederacea (ground ivy) 34
Gonepteryx rhamni (brimstone butterfly) 25

Hedera helix (ivy) 23
Helophilus pendulus (sun fly) 17, 18

Iris pseudacorus (yellow flag) 24-25

Lamium maculatum (spotted deadnettle) 40
Laothoe populi (poplar hawk moth) 32, 33
Lonicera periclymenum (honeysuckle) 33
Lotus corniculatus (bird's-foot trefoil) 20
Lycaena dispar (large copper butterfly) 19
Lycaena icarus (common blue butterfly) 20
Lycaena phlaeas (small copper butterfly) 26
Lythrum salicaria (purple loosestrife) 28

Macroglossa stellatarum (humming bird hawk moth) 32
Meganchile centucularis (patchwork leaf-cutter bee) 20
Melandrium rubrum (red campion) 32
Mellinus arvensis (field digger wasp) 29
Mimas tiliae (lime hawk moth) 32, 33
Muscari racemosum (grape hyacinth) 29
Myosotis arvensis (forget-me-not) 26

Nymphalis io (peacock butterfly) 25

Orchis mascula (early purple orchid) 37
Osmia rufa (red osmia) 34

Papaver rhoeas (poppy) 23
Papilio machaon Britannicus (swallowtail) 28
Parnassia palustris (grass of Parnassus) 14, 30
Pastinaca sativa (wild parsnip) 21

Phyllopertha horticola (garden chafer) 22
Plantago lanceolata (ribwort plantain) 17
Primula veris (cowslip) 29
Psithyrus rupestris (hill cuckoo bee) 27
Psychoda alternata (owl midge) 39

Ranunculus species (buttercup etc.) 36
Rhingia campestris (hoverfly) 24, 25
Rosa spinosissima (burnet rose) 22

Salix caprea (goat willow) 18
Salvia pratensis (meadow sage or clary) 34, 35
Scaeva pyrastri (swarming hoverfly) 30
Scrophularia nodosa (figwort) 29
Silene nutans (Nottingham catchfly) 33
Syrphus balteatus (hoverfly) 30, 41

Syrphus ribesii (currant hoverfly) 30

Thalictrum species (meadow rue etc.) 36
Tilia cordata (lime) 23
Trifolium pratense (red clover) 19
Trifolium repens (white clover) 19
Tussilago farfara (coltsfoot) 17

Vaccinium myrtillus (bilberry) 27
Vanessa atalanta (red admiral) 25
Vanessa cardui (painted lady) 25
Vanessa urticae (small tortoiseshell) 17, 18, 25
Vespa crabro (hornet) 24
Vespa vulgaris (common wasp) 23, 37
Vicia cracca (tufted vetch) 20
Viola tricolor (wild pansy) 31

Index

All numbers in this index refer to pages, not to picture numbers

anemones 36
angelica 21
ants 11
arum lily 39

bean 19
bees 12, 13–14, 15, 41
beetles 14
bilberry 27
bird's-foot trefoil 11, 20, 41, 49
birthwort 38, 41, 46
bisexual flower 9, 10, 52
brimstone butterfly 25
broom 19, 49
buddleia 13, 25, 41
bugs 14

bumblebee, early 19
  red-tailed 17, 18, 27, 35, 41
  small garden 23, 36, 41
bumblebees 13, 14, 15, 18, 19, 23, 34, 35, 36, 41
burnet rose 22
buttercup 36
butterflies 12, 13, 25, 41

carbon dioxide 12
caterpillars 13
celandine, lesser 37
chafers 15, 22, 37
clary (meadow sage) 34, 35
clovers 19, 41, 49
colour change of corolla 27
colour in flowers 12, 27, 31
coltsfoot 17, 41, 44–45
columbine 35, 36

common blue butterfly 20
cornflower 17, 41
cowslip 29, 47
cresses 44
crucifer family 44
cuckoo-pint (lords-and-ladies) 39
currant hoverfly 30, 41

deadnettle 41
delphinium 36

field digger-wasp 29
figwort 29, 48
flies 14
flower bee 29, 41
flower bug 14
flower spider 15
flowering rush 14, 24

forget-me-not 26, 27, 41, 45
fritillary, Queen of Spain 31

garden chafer 15, 22
gnats 14
goat willow 18
gorse 19
grape hyacinth 29
grass of Parnassus 30, 41, 46
green hairstreak 27
ground ivy 34, 41, 50

harlequin fly 38
hemlock 21
hill cuckoo bee 27
honey 14, 15–16
honey-bee 15, 16, 23, 24, 26, 29, 41
honeysuckle 33, 41, 50
hornet 24
hoverflies 14, 18, 24, 25, 30
humming-bird hawk moth 32
hyacinth 29
hybrid 10, 47

ivy 23, 41

labiate flowers 50
laburnum 19
ladybird 14
large copper butterfly 19
larkspur 36
leaf beetle 14, 21
leaf cutter bee 20, 41
lesser celandine 37
lime 13, 23, 41, 43
lime hawk moth 33
ling 26, 41, 45
long-horn beetle 14

long horned eucera 40
lords-and-ladies 14, 39, 41, 46

marsh helleborine 14, 37, 51
marsh marigold 36
meadow rue 36
meadow sage 34, 35
mining bee 18, 41
monkshood 36, 48
moths 32, 41
mustard 44

nectar 12, 13, 15, 41
Nottingham catchfly 33, 41

oil beetle 15
old man's beard 36
orchids 37, 51
owl midge 14, 39

painted lady butterfly 25
pansy 31
papilionaceous flowers 49
parsleys 21
peacock butterfly 25
pollination 9–10, 41–51
    by animals 11–12
    by water 11
    by wind 11, 17
    cross pollination 10, 11, 12, 19, 47
    self pollination 10, 11
poppy 23, 41, 42
poplar hawk moth 32, 33
primrose 41
purple loosestrife 28, 47

radish 44
Ranunculus family 36

red admiral butterfly 25
red campion 32, 41
red osmia 34, 41
ribwort plantain 17
rose chafer 37
rosehips 42
roses 22, 41, 42
rosebay willow herb 27, 48

sage 41 (see also meadow sage)
small copper butterfly 26
soldier beetle 14
spiders 15
spotted deadnettle 40
sun fly 17, 18, 41
swallowtail butterfly 28
swarming hoverfly 30, 41
swede 24, 41, 44
sweet pea 19

tortoiseshell butterfly, small 17, 18
turnip 41

umbellifers 15, 21, 43

vetch, tufted 20, 41, 49
violets 31, 41
viper's bugloss 26, 27, 41

wasps 12, 14, 15, 23, 29, 37
wild arum 39, 46
wild carrot 21, 41, 43
wild parsnip 21, 41, 43
willow 18, 41, 44
wood anemone 36
wool-carder bee 26, 41

yellow flag 24–25, 51